D1456579

NANOTECHNOLOGY
Invisible Machines

SANDY FRITZ

A+

021704

PHOTO CREDITS

Page 4: courtesy Penton Media, Inc. Page 5: courtesy Intel Corporation. Page 7: courtesy Dr. Nelson Max, University of California, Lawrence Livermore Laboratory and the Dept. of Energy. Page 9: courtesy Geoffrey Stewart/Hybrid Medical Animation. Page 10: courtesy NASA. Page 11: courtesy Geoffrey Stewart/Hybrid Medical Animation Page 12: courtesy Geoffrey Stewart/Hybrid Medical Animation. Pages 14-15: courtesy Geoffrey Stewart/Hybrid Medical Animation. Page. 16-17: Copyright © 1999 Tim Fonseca. Page 18 courtesy Geoffrey Stewart/Hybrid Medical Animation. Page 20-21: copyright © 2001 Tim Fonseca. Page 22: Copyright © 1999 TIME Inc, reprinted by permission. Page 23: courtesy Bruce C. Schardt, Purdue University. Page 24: Courtesy Michael L. Roukess/Felice Frankel. Page 26: courtesy Bryan Christie Page 27: copyright © 2000 Tim Fonseca. Page 29: courtesy Bryan Christie. Page 31: courtesy Felice Frankel. Page 32: courtesy Bryan Christie. Page 34: courtesy Slim Films. Page 35: courtesy Yuri Gogotsi/Photo Researchers, Inc. Page 36: courtesy Slim Films. Page 37: courtesy Alfred Pasieka/Photo Researchers, Inc. Page 39: courtesy Deepak Srivastava. Page 40: courtesy Hewlett-Packard Laboratories/Photo Researchers, Inc. Page 42: courtesy Peter Menzel. Page 45: copyright © 2002 by Gina 3Nanogirl2 Miller. Page 45: copyright © 2000 Tim Fonseca.

Published by Smart Apple Media
1980 Lookout Drive, North Mankato, Minnesota, 56003

Produced by Byron Preiss Visual Publications, Inc.
Printed in the United States of America

Copyright © 2003 Byron Preiss Visual Publications

Edited by Howard Zimmerman
Associate editor: Janine Rosado
Design templates by Tom Draper Design Studio
Cover and interior layouts by Gilda Hannah
Cover art by and copyright © 2003 to Tim Fonseca

Library of Congress Cataloging-in-Publication Data

Fritz, Sandy
Nanotechnology: invisible machines / by Sandy Fritz.
p. cm. — (Hot science)
Summary: A discussion of the field of nanotechnology, focusing on general and medical applications.

ISBN 1-58340-368-X

1. Nanotechnology—Juvenile literature. [1. Nanotechnology. 2. Technology.] I. Title. II. Series.
T174.7.F75 2003   620'.5—dc21   2003042783

First Edition

9 8 7 6 5 4 3 2 1

CONTENTS

INTRODUCTION   **Just How Small *Is* the Nanoworld?**   4

CHAPTER 1   **Nanotechnology in Medicine**   7

CHAPTER 2   **Nanoshells and DNA**   13

CHAPTER 3   **The Physics of the Nanoworld**   19

CHAPTER 4   **Making Microprocessor Chips**   24

CHAPTER 5   **From Microprocessors to Nanoprocessors**   28

CHAPTER 6   **Putting the Pieces Together**   34

CHAPTER 7   **The Borders of Science Fact**   41

GLOSSARY   46

INDEX   47

Just How Small *Is* the Nanoworld?

INTRODUCTION

Computers in the early 1950s were hand-built and normally required a technician standing by when they were operated. Data, usually complex math problems, was fed into the machine via punched cards (lower right). The computer's solutions were generated on punched cards.

Many people say that technology has made the world a smaller place. Phone lines crisscross the planet, connecting people thousands of miles away. News that takes place on a small island in the South Pacific can be worldwide headlines in a matter of minutes. Airplanes take passengers from continent to continent in a matter of hours. These kinds of technologies, and many others, have not changed the actual size of the world, but they have made the world seem much smaller.

And now technology itself is getting smaller, but in a literal way.

Shrinking technology is most evident in the world of computers. At the end of World War II, a single computer filled a room the size of a basketball court. It required special refrigeration systems to keep its components cool. An army of technicians remained on alert to replace radio tubes as they burned out and to monitor fragile electronic parts. Today, the average laptop computer has a thousand times more computing power than the earliest computers.

Today's computers are made from microcircuits. Rooms full of large radio tubes have been replaced by etched motherboards and tiny microprocessor chips. But right around the corner are developments that could make today's computers seem as primitive and gigantic as the first pioneering efforts. It's likely that

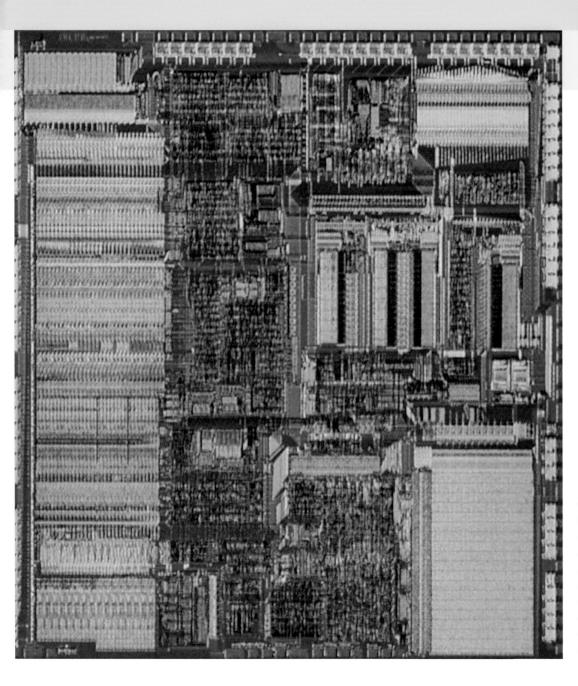

The **Central Processing Unit (CPU)** or "chip" is the heart of all modern computers. It includes more than 100,000 components, or parts, in a 1.5-by-3.5 inch (3x8 cm) area. **Nanosized CPUs** could be thousands of times smaller.

a future generation of computers will not be made from microcircuits but from *nano*circuits.

When people speak of nanotechnology, they are talking about *tiny* in a way that is hard to imagine. In the macroscopic or visible world, a meter is fairly big and a millimeter is fairly small. A millimeter is one-thousandth of a meter. On the microscopic scale, the micrometer is the standard of measurement. A micrometer is one-millionth of a meter. The nanoscale is even smaller. A nanometer is a *billionth* of a meter. That is roughly the size of 10 hydrogen atoms laid side by side, or one-thousandth the length of a typical bacterium. That's tiny.

But the payoffs from the tiny technology could be huge. If some of the experimental work in nanotechnology is successful, it could mean that the entire contents of the Library of Congress could fit on a device the size of a sugar cube.* It could mean the birth of new materials with 10 times the strength of steel but with just a fraction of steel's weight. It could mean delivering life-saving medicine to areas in the body as small as specific molecules.

The possible applications for nanotechnology are limitless. It could change the lives of every person on the planet in ways now hard to imagine. Scientists around the world are investigating, experimenting, and making breakthroughs in one of the biggest areas in science today: the super-tiny world of nanotechnology.

---

*The Library of Congress has more than 18 million books, 12 million photographs, and 54 million manuscripts.

Nanotechnology in Medicine

CHAPTER ONE

Because it is a new field, people have defined the word "nanotechnology" in different ways. But everyone agrees that nanotechnology has three basic properties: Nanostructures range in size from 1 to 100 nanometers. Nanostructures are deliberately designed structures, rather than small particles found in nature. And finally, nanostructures can be combined to form larger structures.

Many natural objects exist on the nanoscale. Most of these nanostructures are found inside the cells of living creatures. The DNA inside a cell's nucleus can be measured in nanometers. And there are important structures called "ribosomes"—

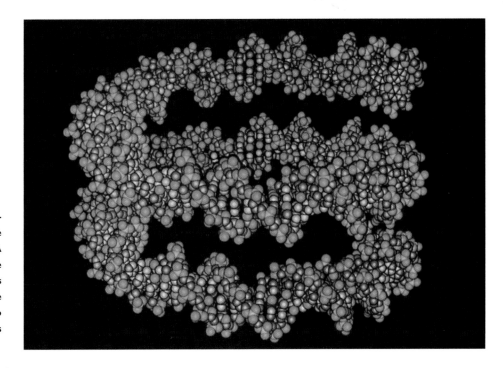

Nature builds elegant structures on the nanoscale. The component parts of the **DNA** molecule, pictured here, are measured in nanometers. Its double-helix shape allows the molecule to fold together so that it can fit into the nucleus of a cell.

tireless factories for producing proteins—that are also nanosized and found in the nuclei of living cells. The nanostructures inside of cells perform important functions. Some break down molecules for the body to use for fuel. Other nanostructures pull molecules apart and process them to provide the raw material for cellular repair and defense.

No scientist working in the realm of nanotechnology thinks she or he can build a machine that would perform better than a living cell. But they do look to nature for inspiration. Studying living cells has helped scientists appreciate the elegance of cellular mechanisms. And in the case of one nanodevice being tested in medical research, nature has provided not only inspiration but the raw material as well.

"Magnetotactic" (magnetic-sensing) bacteria are curious creatures. They live in the muddy bottoms of freshwater creeks and ponds. They are very picky about their environment: They do not like the abundance of oxygen found close to the water's surface. They do not like the scarcity of oxygen found at the bottom of a pond. And yet, these bacteria are so small that gravity has almost no effect on them. So how can these creatures tell up from down when seeking their desired level of oxygen? The answer surprised everyone.

Scientists discovered that these bacteria gain their sense of direction with nanosized magnets. Inside their bodies are 20 nanosized magnetic crystals. Working together, these crystals form a kind of miniature compass. It allows the bacteria to move in response to Earth's magnetic field, which shows them north and south, up and down. When magnetotactic bacteria drift away from their ideal location into zones with either too much oxygen or not enough oxygen, they can flash their whiplike tails and follow Earth's magnetic lines of force back home.

After studying the nanosized natural magnets in these bacteria, scientists have created artificial nanomagnets, and they are using them in a whole new way. They are attaching them to antibodies, which are objects in the bloodstream that are used as defenses against disease and infection. Scientists are using the magnetized antibodies to help detect bacteria and viruses in the human body.

Another nanosized structure called a ribosome (pictured here) is an efficient mechanism. Inside the body, ribosomes combine complex blends of molecules to manufacture the proteins that make all life possible.

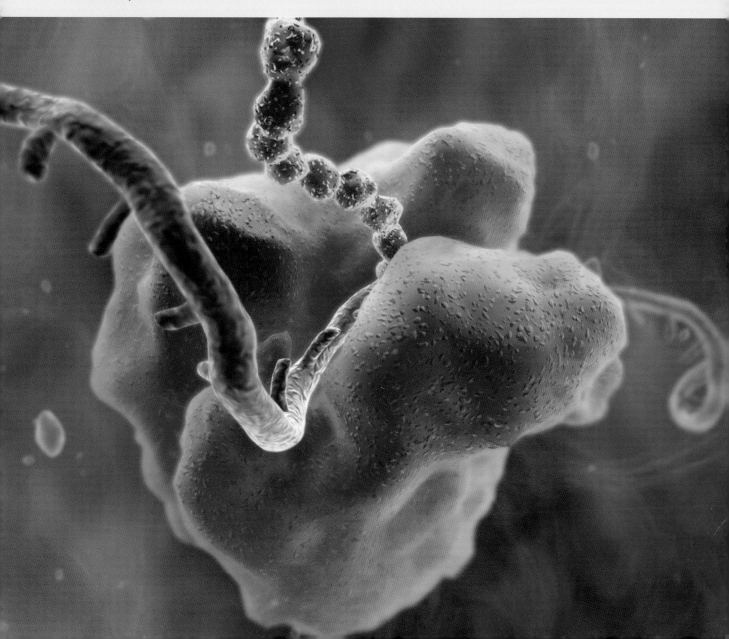

The process, still under development, works something like this: Nanosized magnets are attached or "tagged" to a variety of antibodies. Specific types of antibodies attack specific types of microorganisms. Medical researchers take many different nanomagnet-tagged antibodies and add them to a blood sample from a patient being tested for disease-causing germs.

Antibodies act as keys and germs are like locks. A specific antibody will fit into and attack only a specific germ. When many different kinds of antibodies have been added to a sample of the sick patient's blood, some will link with the germs and some will not. To find out which antibodies have attached to the germs, a strong magnetic field is applied to the sample. Nanomagnetic-tagged antibodies that fail to attach to any germs will line up with the magnetic field. Antibodies that are attached to the germs will not. This makes it possible for doctors to know which germs are making the patient sick and which specific antibodies to use to make the patient well.

The benefits of an approach like this are important. Many times a doctor is not certain which germ is responsible for a person's illness. The treatment prescribed is often based on a best guess. Using the magnetically tagged antibody test elim-

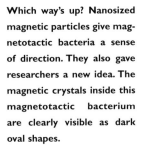

Which way's up? Nanosized magnetic particles give magnetotactic bacteria a sense of direction. They also gave researchers a new idea. The magnetic crystals inside this magnetotactic bacterium are clearly visible as dark oval shapes.

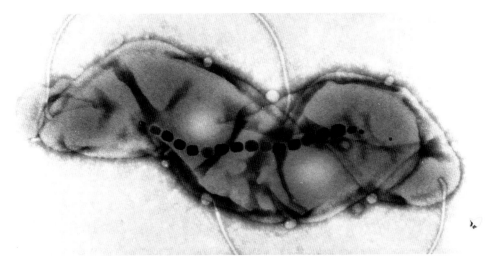

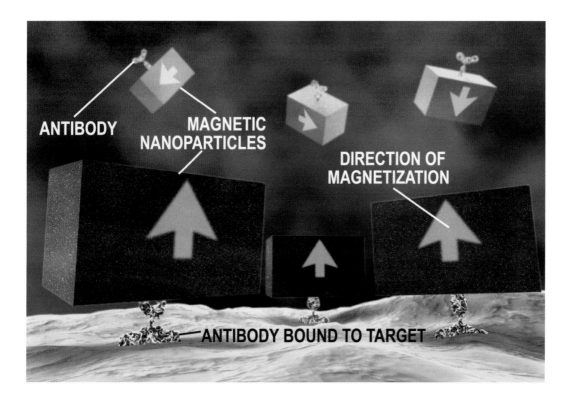

ANTIBODY   MAGNETIC
NANOPARTICLES

DIRECTION OF
MAGNETIZATION

ANTIBODY BOUND TO TARGET

inates the guesswork, allowing patients to receive the perfect treatments in the shortest amount of time.

Another tracking method is under development that uses nanosized beads. Latex beads filled with different colors are linked with, for example, antibodies and released into a blood sample. To see if the antibodies bind with the diseases, researchers shine light on them. Depending on the color returned by the sample, scientists can tell which antibody is binding with which disease. This helps them identify the disease.

Other experiments are underway today that use nanostructures to help boost the power of the body's defenses. Researchers have recognized that newly invented nanosized glass beads coated in a thin layer of gold, called "nanoshells," can be used to deliver molecules of medicine inside the body. Once a positive identifica-

Researchers have used nanosized magnets to pinpoint disease in the body. Antibodies tagged with nanosized magnets are added to a blood sample. By tracing which antibodies link to which bacteria or viruses, doctors can better pinpoint a treatment.

tion of a disease-causing germ has been made, nanoshells containing medicine can be attached to antibodies such as white blood cells. They would then be either swallowed in pill form or injected into the body. The result would be a one-two punch. The antibodies naturally seek out the germs and start their defensive work, while the nanoshells dissolve when they reach their intended target and release medicine to kill the invaders.

Tumors produced by cancer also trigger a response from the body's defenses. For fast-growing tumors, medicine is not always effective. The traditional treatment is to flood the body with medicine, but very little of it arrives at the target tumor. As a result, doctors must frequently resort to radiation treatments that are themselves traumatic to the body. Here is another situation where nanoshells can be helpful. Inside nanoshells that are either swallowed in pill form or injected into the body, modified antibodies can carry their packages of powerful medicine right to the very cells that are tumorous. This kind of treatment pinpoints the troubled area and directs medicine to exactly where it is needed. It is still experimental, but when ready, it may eliminate the use of many other cancer treatments.

Nanoshells could be used to target tumors. Either injected into a tumor or guided there by antibodies, nanoshells could deliver medicine to specific targets. With gentle heat added, they explode, helping destroy the tumor. Nanoshells already in tumorous material (right) are bathed in light from an external source.

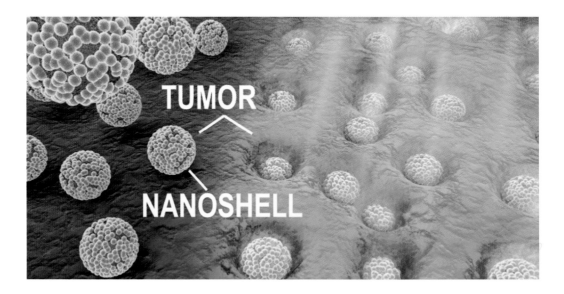

Nanoshells and DNA

Nanoshells may have many interesting applications in future medical science. One idea being explored by medical scientists is using nanoshells as tiny "bombs" that destroy tumors on contact. In this scenario, nanoshells are simply injected into that part of a patient's body where a tumor has been identified. When they are lodged in or near the tumor, a warm lamp is placed on the skin above the area. The gentle heat causes the nanoshells to rupture. They burst open, delivering their cancer-fighting medicine right where the tumor is. And the tiny explosions help damage the tumor while leaving nearby tissue unharmed. Delivering medicine via nanoshells seems to have a promising future.

Another use for gold-coated nanoshells is to test for whether or not a person has been born with a genetic disease by testing a sample of the person's DNA. DNA is the substance in the nucleus of each living cell that carries that organism's complete genetic code. In the DNA are coded instructions for everything from the person's eye color and foot size to intelligence and disease. In some cases, knowing that your DNA carries the code for a certain disease can lead to early treatment or preventative measures.

This test exploits the unique characteristics of DNA. Incomplete DNA strands have a natural tendency to complete their sequences by binding to the missing pieces of the strands and forming stable structures. Researchers have taken short segments of DNA (incomplete strands) and attached them to gold nanoshells. The modified nanoshells are introduced to a sample that is being test-ed for a specific genetic sequence (a genetic disease). During the test, the frag-ments of DNA on the nanoshells are drawn to certain genetic sequences in the sample and bind with them. As more and more modified DNA binds with

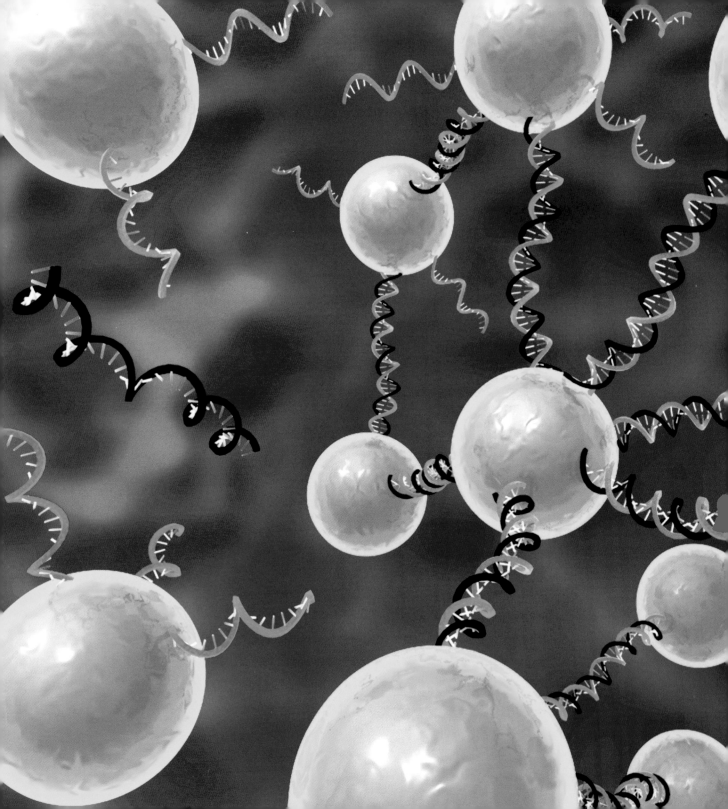

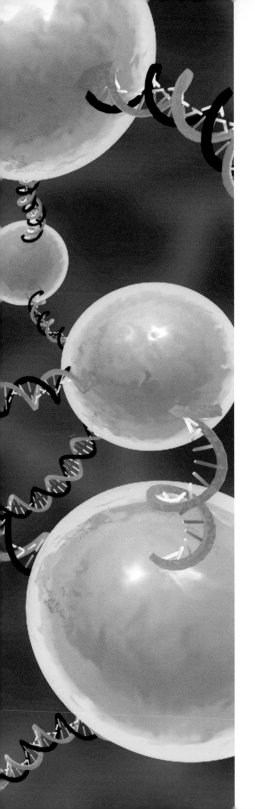

sequences from the patient's blood sample, more and more gold balls are trapped, too. When enough of these gold nanoshells bind with specific DNA sequences in the solution, the sample changes color. This color change shows the doctors that the sequence they were searching for is present. If this happens, they can then tell the patient that he or she does carry the genes for a certain disease, long before that disease ever appears.

Still another idea also uses the natural tendency of DNA to complete itself and form stable structures. In this case, a nano-sized beam or lever is covered with incomplete sequences of DNA. When a sample being tested is introduced to the DNA-enhanced beams, some of the sequences bind to it. The binding action actually bends the tiny beam, sending a signal to researchers that certain sequences of DNA are present in the test sample. This way, researchers can discover if the genetic sequences they are looking for are present in the sample.

Many of the innovative uses for nanotechnology in medical treatment fall into the same category: Good ideas that need much more work before they are ready to be tried on human beings. Another such idea is to use nanosized building blocks to help enhance the healing of broken bones or of damaged tissue. Synthetic molecules, assembled piece by piece at the atomic level, have shown promise in building structures on which new bone can grow. When targeted to broken bones or damaged

Gold nanoshells covered with short strands of DNA tend to connect with other DNA strands. When they match, scientists can look for those genetic sequences that are specific to known diseases.

Robots in the Bloodstream

In the distant future, medical scientists hope to construct "nanosubmarines" that could patrol the human body performing hunter-killer functions. Onboard sensors would detect any kind of abnormal conditions. The nanosubmarines would diagnose the problems with onboard laboratories. Then, depending on their abilities, they would either treat the condition themselves or activate the body's own immune system for help. The benefit of the nanosubmarine is that it could detect serious illness—such as cancer—long before the problem became bad enough to trigger the body's natural defensive mechanisms.

An artist's concept of a nanosubmarine in a person's bloodstream, patrolling for unwanted viruses and bacteria.

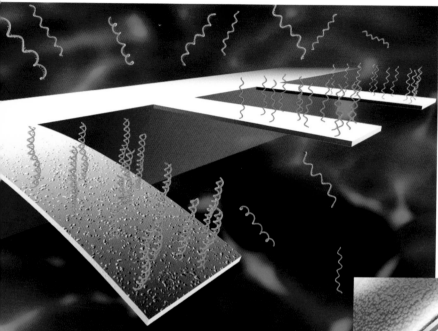

Nanosized beams covered with DNA can attract complementary DNA strands. The weight of the DNA causes the beam to bend, allowing scientists to detect its presence.

To help strengthen the connection between artificial implants (left) and bone (right), nanotechnology may be useful. Complex nets made of nanoparticles can give healing bone and cartilage a firmer base upon which to build new tissue.

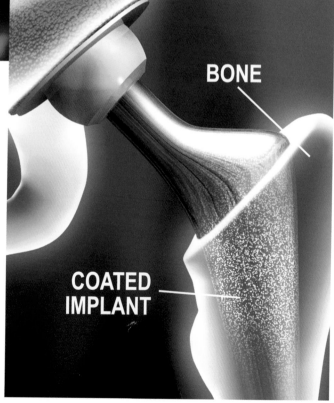

BONE

COATED IMPLANT

cartilage, the shape and texture of the tiny molecule actually attracts new cells and promotes their growth. With the nanostructures in place, the growth of new cells in the damaged area would be much speedier than if the body had to do the work on its own.

Compared to the marvels of the structure and function of living cells, medical science's first steps in the nanotechnology fields seem both low-tech and humble. But these first steps could have not only immediate benefits, they could also provide the experience for future, more radical developments.

The Physics of the Nanoworld

In many ways, the nanoworld is kind of a "twilight zone." Objects in the nanoworld are far smaller than the objects that make up the world we live in, the macroworld. A paper clip, a pen, or a rubber band are composed of trillions and trillions of atoms. The collective behavior of trillions of atoms clumped together gives macroworld objects their specific characteristics. But the nanoworld is the realm of small numbers of atoms and molecules. At this scale, objects behave much differently than they do in the macroworld. If any kind of manufacturing is to take place on the nanosized scale, scientists are going to have to come to grips with the curious physics of this world.

Quantum physics is a set of theories that describes the behavior of individual atoms and the even smaller particles that make up atoms. The behavior of atoms, according to quantum physics, defies our intuitive understanding of the world. For instance, on the super-tiny scale, subatomic particles can randomly pop in and out of physical existence. Sometimes they appear as particles, but sometimes they appear as energy waves. Another strange aspect of quantum physics: Nothing exists until it is observed. Before being observed or acted upon, the particles that make up physical matter exist only as waves of probability.

Fortunately, proposed nanodevices are not composed of just electrons or a single atom but groups of atoms ranging from one to a few hundred nanometers in size. Therefore, the laws of quantum physics do not directly apply to nanodevices. But nanodevices are not completely free of their effects. The laws of classical physics, those that govern our familiar macroworld, do not fully apply to nanodevices either. Scientists have coined the word "mesoscale" to describe the physics of the nanoworld. Mesoscale is an in-between state, not quite the ultratiny world

Nanopower

Nanosized devices will not be able to plug into a wall socket or run on double A batteries. So where will they get their power from? It depends on the application. If nanosized machines are inserted into the body, scientists hope to be able to have them run on the same fuels that cells use—sugar and oxygen. Batteries are an option for some applications. Inside nanosized devices, these energy storage devices would probably look nothing like their macroworld counterparts, but they would perform the same function. Another source of energy being considered for nanosized devices is solar energy. Some devices may use the voltage found in sunbeams to power themselves. Others may use the photosynthesis mechanism found in plants as a source of power. The problem of how to power nanosized devices is also an opportunity for scientists to explore some unconventional power sources.

Powering nanosized devices is a challenge. Inside the body, nanomachines may use the same energy source used by cells—complex sugar molecules. Pictured here are nanorobots that attach to the inside walls of blood vessels and release nanoprobes (seen as green spheres) to cleanse the blood.

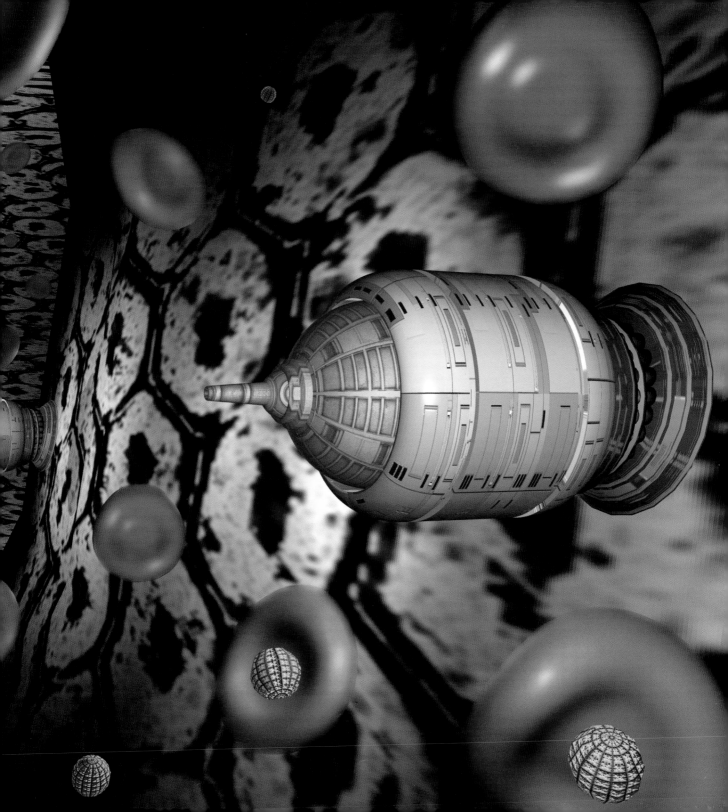

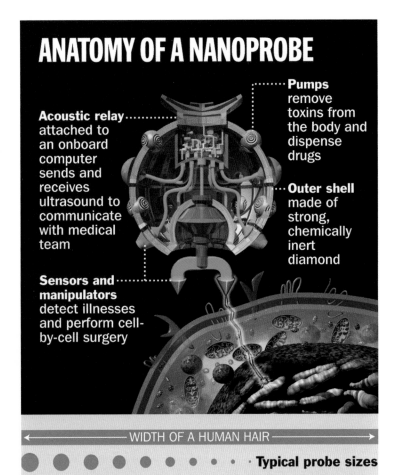

ANATOMY OF A NANOPROBE

Acoustic relay attached to an onboard computer sends and receives ultrasound to communicate with medical team

Pumps remove toxins from the body and dispense drugs

Outer shell made of strong, chemically inert diamond

Sensors and manipulators detect illnesses and perform cell-by-cell surgery

◀─────── WIDTH OF A HUMAN HAIR ───────▶

● ● ● ● ● ● ● ● · **Typical probe sizes**

Up to 10 trillion nanorobots, each as small as 1/200th the width of a human hair, might be injected at once

Researchers foresee a day when nanoprobes 1/200th the width of a human hair patrol the body. This model includes a relay to communicate with an outside computer, arms that could be used for nanosurgery, and pumps to dispense drugs. (Note the scale given at the bottom in the red horizontal bar.)

where quantum physics rules, but not quite a part of our classical physics macroworld. Scientists are just beginning to understand mesoscale physics.

The ongoing study of this new arena blends the skills of the chemist, the physicist, and the engineer. One of the early findings in mesoscale physics represents a big hurdle: The smaller the device, the more susceptible it is to alteration from outside forces. For example, scientists would like to make a nanosized scale to weigh single atoms. This simple idea shows how the physics of the mesoworld must be considered when making nanodevices.

Atoms are not stationary objects. They vibrate. Groups of atoms form molecules, and molecules vibrate too. Placing an atom to be weighed on a nanosized scale actually changes the physical characteristics of the weighing device itself. The atom being weighed vibrates and interacts with the scale's own vibrations. This interaction throws off the scale's ability to measure weight. Even with no sample to weigh, the properties of the scale shift continuously. The atoms and molecules of the weighing device absorb and release electrons and atoms. The result is random instability. Scientists hope to work around this problem by using large numbers of weighing devices to average out the fluctuations. But this kind of problem represents the challenges faced by scientists working with physics at the new mesoscale.

Another key problem that has appeared on the mesoscale of nanodevices is providing them with power. The power needed to operate nanodevices would be a millionth to a billionth of the power needed to run a conventional computer. This kind of power draw would be a blessing for electronics users—a single, tiny battery could keep a nanodevice running for a long, long time. But creating power levels on this tiny scale has never been attempted before. Doing it will take a totally new approach to designing a power source.

Understanding the physics of the mesoscale environment in which nanotech devices will operate is crucial if the devices are to be as sturdy and useful as present-day computers.

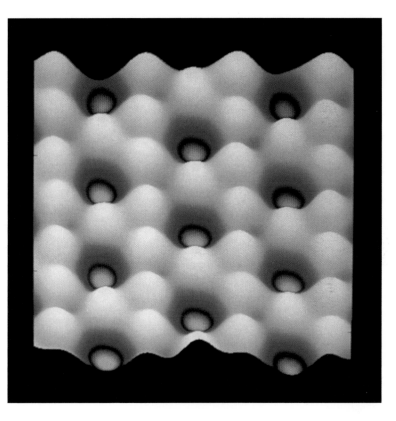

A scanning tunneling micrograph (STM) image of actual iodine atoms on a platinum surface. Being able to see atoms at this scale helps researchers fashion nanomachines.

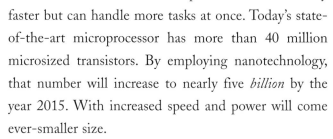

Making Microprocessor Chips

CHAPTER FOUR

Big enough to avoid many of the problems found in quantum physics, yet far smaller than anything ever manufactured before, nanosized objects occupy a little known physical environment called the mesoscale. Pictured here are experimental nanoresonators. Resonators are used in a variety of electronic machines.

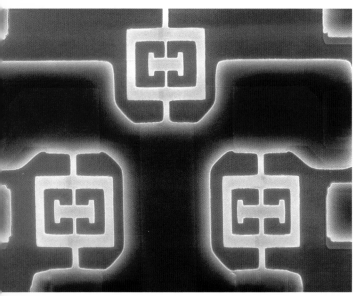

Most people predict that nanotechnology's biggest impact will be in computers. Computers are constantly getting faster. And yet computer microprocessors—the "brains" of any computer—are close to the limit of how small and fast they can become while still using traditional methods. Within five years, computer microprocessors must become *nano*processors if the speed and usefulness of computers is to continue to grow.

The speed and processing power of a microprocessor is based on the number of transistors it has. Transistors are basically on/off switches that direct the flow of electricity through a circuit. The more transistors, the more speed and power a microprocessor has. More transistors translate into computers that are not only faster but can handle more tasks at once. Today's state-of-the-art microprocessor has more than 40 million microsized transistors. By employing nanotechnology, that number will increase to nearly five *billion* by the year 2015. With increased speed and power will come ever-smaller size.

Making nanosized computer processors is a tricky job. To understand the problems of making increasingly smaller processors, it's important to understand how today's microprocessors are made.

Most manufacturers of microcircuitry use a technique called "photolithography" to produce microprocessor chips. This process uses light (photo) to etch out (lithography) the patterns of a microprocessor. The first step is to create a large version of what will eventually become the micro-processor, so an engineer begins by designing a microprocessor on a large scale.

When the design of the microprocessor is approved, the pattern is transferred onto a plate that will serve as the master pattern, called a mask. This is done by etching the complex pattern into the plate using a laser beam. The beam transfers an exact copy of the circuit design onto the material, creating the "mask" or tem-plate that will be used to produce the actual microprocessor.

But the mask is far larger than the microprocessor it will ultimately produce. To shrink the mask's size, it is fitted into a machine with a lens. The lens takes the large design and shrinks it down into a much smaller size. With the mask in place and its pattern shrunk down, ultraviolet light is directed through the mask. The solid parts of the mask block the light, while the exposed areas allow the light to shine through. The result is an exact miniaturized version of the pattern from the template, now etched into a two-layered sandwich made of silicon. A silicon microprocessor chip is born. (Many people shorten the name and refer to this object as a "microchip.")

Using this technique, tiny channels measuring down to 100 nanometers across can be carved into the silicon wafer. The resulting microprocessor is about one-inch (2.5 cm) square. Today's best microprocessors pack over 40 million tiny com-ponents into a square this size.

A computer "brain" measuring one inch (2.5 cm) square is pretty small. And yet, scientists working with nanotechnology want to go much smaller. Instead of channels and grooves measuring 100 nanometers, they want to build processors with channels and grooves that measure around 20 nanometers.

PHOTOLITHOGRAPHY: HOW COMPUTER MICROCHIPS ARE MADE IN 2003

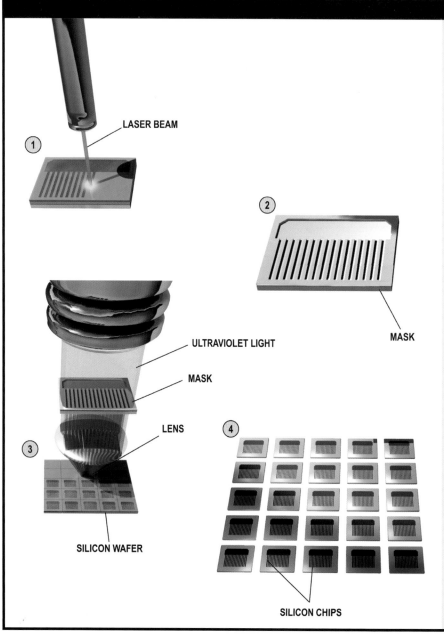

1   LASER BEAM

2   MASK

3   ULTRAVIOLET LIGHT   MASK   LENS   SILICON WAFER

4   SILICON CHIPS

PHOTOLITHOGRAPHY USES LIGHT TO ETCH MICROCIRCUITS ONTO SILICON WAFERS.

1. A laser etches a pattern onto a plate.

2. When cleaned, the plate is called the mask.

3. The mask is fitted into a machine that shines a beam of ultraviolet light through the mask. Parts of the light are blocked by the mask, others go through. The light that makes it through the mask is focused by a lens to etch out a circuit pattern onto a silicon wafer.

4. The end result is rows of identical microchips.

Moore's Law

Another version of a robotic nanoprobe. This one includes arms with gripping hands and an energy beam to zap disease-causing bodies in the blood stream.

In 1965, Intel co-founder Gordon Moore made a prediction: The number of transistors on integrated circuits, such as computer microchips, would double every 12 months. At first, the prediction held true, but a slight slowdown in growth made Moore modify his prediction to a doubling every 18 months. The prediction, known as Moore's Law, has held true for more than 25 years. If computer microprocessors continue following Moore's Law, nanosized computer processors should first appear between 2010 and 2015.

From Microprocessors to Nanoprocessors

CHAPTER FIVE

The first method used to create nanosized computer components took advantage of the manufacturing techniques common in making microprocessors. This is called a "top-down" manufacturing technique. Much research has been directed toward top-down methods. One important aspect of simply shrinking current manufacturing techniques is that the existing techniques are familiar to microchip makers. Another consideration is that a successful top-down technique would save money. It would allow manufacturers to modify their current machines rather than having to replace them. This would save billions of dollars.

To make circuits ever smaller, scientists have experimented with replacing the ultraviolet beam used to carve patterns in the photolithography method with an electron beam. An electron beam is much more precise than ultraviolet light and allows for the cutting of even smaller channels and grooves into a processor chip.

Electron-beam lithography allows researchers to create patterns measuring only a few nanometers wide. But using this technique has some drawbacks. Although the process is familiar, it will require a host of new, and quite expensive, machines. More important, the process is very slow. The electron beam is so concentrated and so tiny that it takes a long time to create the processor pattern. Creating a computer processor with electron-beam lithography is like coloring over the floor of a basketball court with a single crayon. Sadly, it would be next to impossible to mass produce nanosized processor chips through the use of electron-beam lithography.

X-rays have also been tried in the lithography process, but they have their own set of problems. X-rays cannot be focused by lenses the way ultraviolet light can, and X-rays tend to damage the mask, the lens, and the silicon chip beneath.

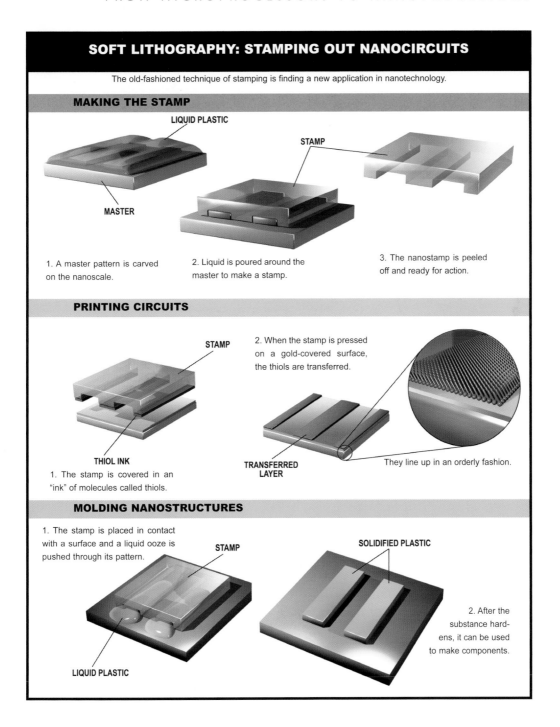

SOFT LITHOGRAPHY: STAMPING OUT NANOCIRCUITS

The old-fashioned technique of stamping is finding a new application in nanotechnology.

MAKING THE STAMP

LIQUID PLASTIC

STAMP

MASTER

1. A master pattern is carved on the nanoscale.

2. Liquid is poured around the master to make a stamp.

3. The nanostamp is peeled off and ready for action.

PRINTING CIRCUITS

STAMP

2. When the stamp is pressed on a gold-covered surface, the thiols are transferred.

THIOL INK

TRANSFERRED LAYER

They line up in an orderly fashion.

1. The stamp is covered in an "ink" of molecules called thiols.

MOLDING NANOSTRUCTURES

1. The stamp is placed in contact with a surface and a liquid ooze is pushed through its pattern.

STAMP

SOLIDIFIED PLASTIC

2. After the substance hardens, it can be used to make components.

LIQUID PLASTIC

Frustrated by the problems of trying to shrink down traditional microcircuitry techniques, researchers began brainstorming new ideas. Surprisingly, one of the most promising techniques for producing nanosized circuitry is quite low tech.

This simple process is called "soft lithography." It is a process that is similar to making a rubber stamp. It starts with an electron beam carving a circuit pattern into a piece of silicon. This piece becomes the master template. Then, scientists pour a rubbery fluid over the master pattern. When it hardens, they peel it off. The hardened fluid now has a raised pattern of the circuit on its underside that matches the original. Amazingly, details as fine as only a few nanometers are faithfully reproduced in the hardened rubber stamp.

The soft lithography stamp is used much like an ordinary rubber stamp. To make nanosized computer components, the stamp is covered with a solution of organic molecules called "thiols." Thiols are used because they line up in ordered patterns automatically, the way a compass needle automatically points to the north. When the stamp covered in the thiol solution is brought into contact with a thin layer of gold that sits atop a silicon chip, the thiol molecules are transferred to the surface and stick there. The highly ordered pattern is an exact duplicate of the original stamp. This technique requires no special equipment and can be carried out by hand in most laboratories.

But soft lithography is not quite the breakthrough that is needed to catapult nanoprocessors into computers by next Christmas. Current microprocessors are complex, stacked layers of interconnected circuitry. The soft lithography method, while fast and easy, is not as accurate as it needs to be to ensure that multiple layers of circuit patterns line up exactly. Even the tiniest flaw would render the component useless. Experiments with stamps created from ultra-hard, ultra-pure crystals have had better success, but still more development is needed before this method has a chance to produce nanosized processor chips.

The final technique for nanofabrication that shows real promise relies on a breakthrough invention—the atomic force microscope (AFM). It operates like an

The "business end" of an atomic force microscope. Its point is so
fine that it can actually trace the outlines of individual atoms. It
can also be used to trace circuit patterns on nanosized scales.

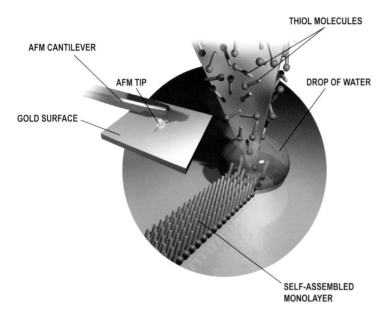

AFM CANTILEVER

AFM TIP

GOLD SURFACE

THIOL MOLECULES

DROP OF WATER

SELF-ASSEMBLED MONOLAYER

Techniques such as dip-pen lithography allow complex patterns to be transferred onto nanosized surfaces. This diagram of the process shows how nanocircuits are "drawn" with a pen. The smaller image inset on the left shows the tip of an atomic force microscope (AFM) tracing a pattern onto a gold surface. Blown-up in the larger circle is the same image. Here you can see the tip of the AFM that has been coated with a solution of thiol molecules. And the drop of water is also visible. This is used by the molecules as a bridge between the tip of the pen and the gold surface. As the pen traces its pattern, the molecules follow, leaving a line of "self-assembled monolayer" behind. This is the pattern that will be traced to create a mask. A nanostamp will then be made from the mask.

old-fashioned record player. The device is so sensitive that it can actually view and identify individual atoms. The AFM works by bringing a tiny probe into direct contact with the sample under investigation—be it a single atom, a molecule, or a germ. As the top of the probe—typically between 2 and 30 nanometers wide—drags across the sample, it bends. The probe follows the shape of the sample's surface, sending its data to a computer. The computer builds up an image of the object line by line.

Nanotech pioneers have developed ways to make the AFM more than just a way to view nanosized objects. The tip can be used to drag molecules and arrange them into complex patterns on a surface. It can also make nanosized scratches on a surface. And if electrical current is directed through the tip, the atomic force microscope can become a miniature electron beam source. This beam, in turn, can etch nanosized patterns into a surface.

A promising new use of the AFM is called "dip-pen lithography." The tip of the AFM is dipped in a solution of thiol molecules, the same as those used in soft lithography. The tip is brought near a thin sheet of gold but never actually touches it. Molecules of water form between the AFM's tip and the gold plate. Thiol molecules travel across the water bridge and are deposited on the gold, where they stick. Using this procedure, researchers have created nanolines only a few nanometers wide.

One great advantage to the dip-pen lithography method is that it works with a variety of solutions, not just thiols. It can also be used to write lines directly on silicon. But at this stage, dip-pen lithography is also a slow method. Researchers have not yet determined the best applications for the technique. It may be useful for the precise modification of circuit boards, which are devices that arrange transistors and direct the flow of electricity throughout the maze of components that allow a computer to operate.

Nanofabrication methods such as dip-pen lithography, soft lithography, and electron-beam lithography are not yet polished techniques. They are active experiments. Exploration into these and other techniques will no doubt provide important tools as we move into the coming age of nanotechnology.

Seeing the Nanoworld

Research work at IBM in the early 1980s led to one of the most important tools for nanotechnology researchers—the atomic force microscope (AFM). In 1981, IBM researchers invented the scanning tunneling microscope (STM). The device was a breakthrough—it allowed scientists for the first time to "see" things on the atomic level. It did this by sending a small electric charge through the sample under study. As the charge flowed through the object, a computer connected to the STM was able to construct an image of it. But the STM's abilities were limited. It only worked with materials that could conduct electricity.

This limitation led to the invention of the AFM. Instead of sending a charge through the sample, the AFM dragged the tip of a tiny, sensitive probe across its surface. Computers turned the up and down movements of the probe into images. The AFM has become one of the most important tools for nanotechnology researchers. It allows them not only to see atoms but also to move them around.

Putting the Pieces Together

CHAPTER SIX

A chemical reaction that adds a single electron to this ben-zenethiol molecule (right), causes it to conduct electricity like a tiny wire. Subtract the electron and the switch is "off," meaning that it is non-conducting. The ability to alternately change this molecule from conducting to non-conducting, and back, makes it a candidate for use as a nanosized transistor.

No one doubts that computers will be revolutionized by nanotechnology. The goal is to make processors not only far smaller than today's microprocessors but also far more powerful. As in all fields where nanotechnology is being explored, computer manufacturing using nanosized devices is still in an early stage.

And yet, certain components found in today's computers must have nanosized equivalents if nanoprocessor-based computers are to become a reality. The basic building blocks of computers have, in fact, already been manufactured on the nanoscale. The challenge facing researchers is putting all the pieces together into something that works.

In today's microelectronics, silicon is used as a "semi-conducting" transistor. This is simply a switch that can be turned on or off, flipping between a conducting state (on) or a nonconducting state (off). Transistors are at the heart of a computer processor's ability to function.

In a nanosized computer system, one promising candidate for taking the place of silicon as a semiconducting transistor is a molecule. Recent work has demonstrated that clusters of molecules can take the place of silicon and function as transistors in a nanosized computer system. A chemical reaction between the molecules fuels the semiconducting behavior. The chemical reaction imparts an extra electron to the molecular matrix, causing it to flip to an "on" position. Another chemical reaction can remove the extra electron and block the electron

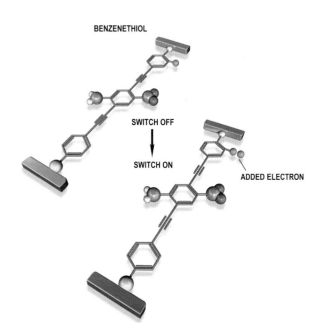

BENZENETHIOL

SWITCH OFF

SWITCH ON

ADDED ELECTRON

flow the way folding a garden hose in half blocks the flow of water. Certain molecules, therefore, are a potential replacement for silicon transistors in a nanocomputer.

Another promising material for replacing silicon are structures called "carbon nanotubes." These interesting structures are collections of super-tiny carbon molecules that shape themselves into pipelike pieces. The exciting thing about these structures is that carbon nanotubes can perform multiple functions. Nanotubes can be created as wires to guide electron flow between a computer's components. And they can act as transistors, regulating the electron flow with on-and-off states. When nanotubes are made of a combination of metal and carbon, they can also be coaxed into acting as "diodes," which are devices that channel the flow of electrons in one direction.

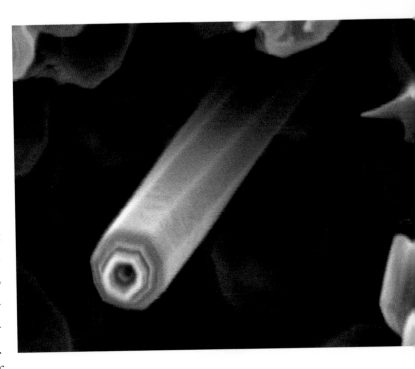

This carbon nanotube is a self-assembling form that is grown rather than manufactured. Carbon nanotubes represent the "bottom-up" method of making nanostructures, where construction starts on the atomic level.

One interesting aspect of nanotubes is that they are not manufactured in a traditional way but are actually grown. The technique of starting with raw molecules and coaxing them into useful components is called a "bottom-up" approach. The bottom-up approach to manufacturing nanosized devices is gradually being recognized by research teams as the preferred method of making nanodevices. It requires a good understanding of chemistry and physics to create structures with the bottom-up method. But once mastered, the technique allows molecules to self-assemble into useable components.

Unfortunately, scientists have realized that these structures cannot solve all of the challenges facing nanosized computers. Some carbon nanotubes are strictly conductive—that is, they are always in the "on" position and cannot be turned "off" to produce a workable transistor. In addition, carbon nanotubes vary in size. It would require the production of huge batches of carbon nanotubes in order to isolate the portions that are semi-conductive and of equal length. This problem translates into a large and tedious investment of time in creating techniques for isolating and harvesting the correct nanotubes from a raw batch of the material.

Connecting different components on a nanosized circuit board could be done with nanotubes. The tubes themselves not only carry electricity, but also function as on-off switches in the circuit.

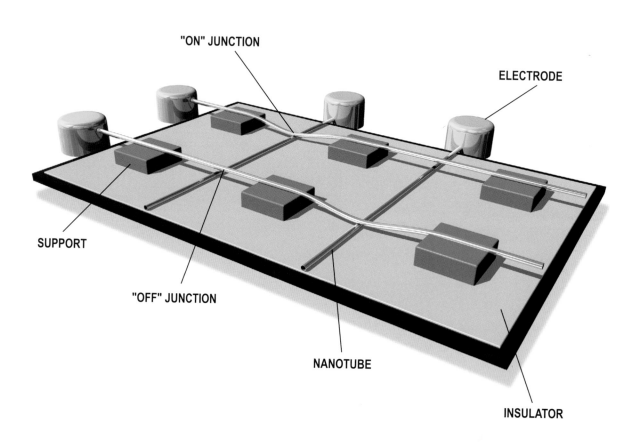

"ON" JUNCTION

ELECTRODE

SUPPORT

"OFF" JUNCTION

NANOTUBE

INSULATOR

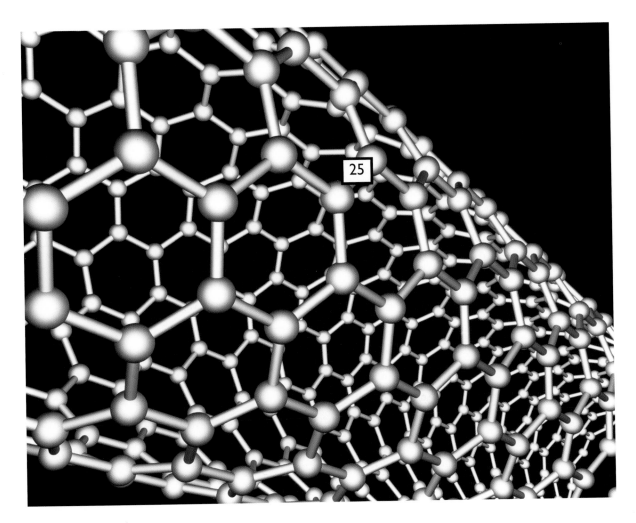

25

Even if a computer helped with the sorting process, it would be a wasteful and time-consuming process. But the idea of a self-assembling product with the qualities sought by researchers is exciting everyone in the field of nanotechnology. Scientists have not given up on carbon nanotubes. And by experimenting with creating different kinds of nanotubes, they may find the answer to the problem.

Knowing which qualities they desire in a nanotube, scientists are also looking at chemistry for the answer. If certain molecules that are semiconducting by

The molecular lattice of a carbon nanotube shows how stable it is. Researchers hope to manipulate some of the chemical bonds in the lattice to make nanotubes of more consistent shapes and sizes.

nature were brought together, perhaps they could be coaxed to grow in a way that would produce the uniform end product scientists are looking for.

Following this line of thought, scientists have succeeded in producing new nanotubes that are both uniform and semiconductive, making them ideal transistors. These manmade nanotubes can also take the place of many critical components inside a computer. They can act as diodes, allowing electrons to flow in one direction only. They also can amplify signals, a necessity to keep electrons flowing smoothly inside a processor. And when these new semiconducting nanotubes are crisscrossed with nonconducting nanotubes, they can trap electrons. Suspending the flow of electrons to hold their positions translates into computer memory. Ordinary computers must refresh their memory once every tenth of a second, and unless the data is saved to the hard drive, it vanishes. In early experiments with nanosized computer memory, scientists have been able to store information for 10 minutes or longer before the memory needs to be refreshed. This type of performance is nothing short of astounding and hints at the kind of breakthroughs that are promised by nanotechnology.

The diverse components needed to create a nanocomputer have each been created and are still being refined. But somehow, the components must be connected together. Early efforts connected components with a top-down technique using metal wires a few nanometers wide. This was quickly replaced by a bottom-up approach using nanotube wires that serve many purposes at once. And yet, computer systems need a maze of interconnections to work. Nanotubes simply do not yet have the kind of flexibility to comprise all the wiring in a nanocomputer. And another problem must be solved: How can the diverse and complex interconnections of a computer circuit board be made on the nanoscale?

Once again, stepping outside the problem and returning to basics has helped supply inspiration for researchers. Picture a drop of coffee on a piece of white paper. If you tilt the paper toward you, the liquid coffee will start to run down to you. If you pick up the left side of the paper, the coffee flows to the right. If you tilt the edge of the paper closest to you upward, the coffee runs downhill away from you. Now picture a nano-sized circuit board etched with tiny channels and apply the same idea. A drop of liquid metal is added to the board and the board is tilted. The liquid metal stays in the channel and flows along the predetermined path. To make a sharp turn, the entire board is tilted, and the metal rounds the turn and continues following its path. By repeating this turning and tilting action of the board, and by allowing gravity to draw the liquid metal through the channels, nanosized wire links between various components can be formed.

The process is actually a bit more complex than it sounds, but it is a great advance. It also allows for complex patterns of nanowires to be layered and stacked atop one another, the same way as in their larger counterpart, the microprocessor.

Nanotechnology has promised to take computing to places no microcomputer has ever been before. And already, in this early stage of development, it has shown astonishing breakthroughs. When a nanocomputer finally does appear, the jump in performance, the tiny size, and the enormous flexibility of its design could render all previous computers extinct.

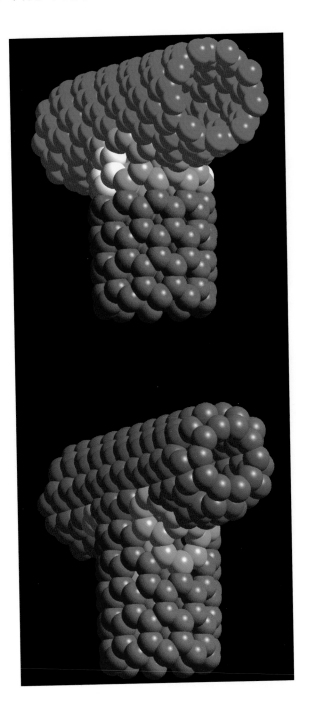

A model showing two nanotubes touching to make a T-shape. Junctions like this could be used to direct the flow of electricity through a nanosized circuit board in a nanocomputer.

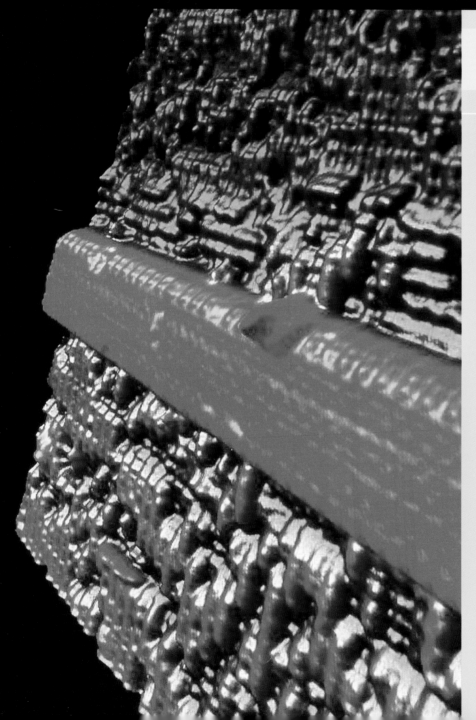

Nanotubes in Space

Carbon nanotubes may have applications in future generations of computers. But they may also have applications in NASA's launch vehicles. Well before the loss of the space shuttle *Columbia* in 2003, scientists had been investigating the use of mats of carbon nanotubes to replace the shuttle fleet's thermal insulation tiles. The benefits are clear—mats of carbon nanotubes are as hard as diamond but far less brittle. The heat-insulation qualities of carbon nanotubes are equal to those of the ceramic material currently used for the shuttle fleet's protective tiles. And the mats weigh significantly less than ceramic tiles.

A scanning tunneling micrograph (STM) of a nanosized wire just 10 atoms wide. Being able to manufacture nanosized components consistently and of high quality is necessary to make nanosized computers a reality.

The Borders of Science Fact

CHAPTER SEVEN

In 1986, a book was published that lit a fire under the scientific community. It was called *Engines of Creation: The Coming Era of Nanotechnology*. Its author, K. Eric Drexler, described a vision of nanotechnology that seemed almost godlike at the time. Nanosized machines, he said, could be injected into the body, destroying cancer cells, heading off any medical ills, and actually reversing the process of aging. Hungry? Push a button, and nanomachines would assemble random molecules into whatever food you desired. Need a car? A watch? A comb? Push a button and allow nanomachines to manufacture these objects in a jiffy. The cost? Next to nothing—the raw material for these objects could be drawn from the soil, from the air, or from the garbage we now throw away.

Drexler did more than just propose a sequence of science-fiction-like miracles. He actually detailed the devices that would make them happen. At the core of his devices was an intriguing machine called a "nanoassembler."

A robotic nanoassembler would work something like this: Picture a robot stationed next to a conveyor belt. As pieces and parts flow past, the robot's multijointed manipulator arm reaches out and plucks parts off the belt. It places the pieces into new patterns with great precision, fixes them in place, and moves to its next task. Now imagine that device shrunk to a scale where it is handling atoms and molecules. This is the nanoassembler, the device at the heart of Drexler's vision of what nanotechnology could be.

Drexler didn't stop there. He envisioned his nanomachines with the ability to reproduce themselves. Connected to a computer rooted in the macroworld, billions and billions of nanomachines working together would be able to assemble, from basic molecules, any object currently produced by macrosized machines. Not only

K. Eric Drexler envisions nanomachines that include gears, ball bearings, and parts familiar in our scale of the world. Devices called "universal assemblers" would make everything we need by assembling atoms into any desired object or substance.

that, but with the ability to manipulate and rearrange atoms in any way, nearly all ills that challenge the human family, from diseases to air pollution, could be set right by nanomachines.

In general, the scientific community scoffed at Drexler's vision. Legions of scientists working in the fields of nanotechnology leveled harsh criticism. No one can even make a *macro*machine that can reproduce itself, never mind fashioning a *nano*machine that could. Another criticism brought against the visionary nanoassembler was that its very substance, atoms and molecules, would inevitably interact with the atoms and molecules it was trying to assemble, rendering it useless. Even if this problem were overcome, how would a nanoassembler break the bonds between molecules to extract the necessary atoms for building? What would it use for power? Criticism went even further. Assuming a nanoassembler could be made, and that it could reproduce itself, what is to stop it from being misused? Perhaps some computer hacker would develop a mutant strain of the nanoassembler that would ignore input from the outside world. Such a device might quickly turn every single atom and molecule on the planet into more nanoassembler devices. It would mean the end of the world as we know it.

Drexler remained indifferent to the criticisms. With money from a wealthy backer, he formed his own company, Zyvex, with the intention of showing the world and his critics that his ideas held merit. Some $20 million later, Zyvex has recognized that the carefully designed, computer-generated blueprints for its nanoassembler will need some major adjustments. For one, building new structures atom by atom just isn't working. Pushing around molecules seems to be easier. But even here, the interactions between molecules at this level of smallness are far more powerful than Drexler and his team originally predicted. A rough version of the nanoassembler has been made, but currently it is 1,000 times bigger than the device needed to begin building on a nanoscale.

Zyvex may eventually make a true nanoassembler similar to the ones envisioned by Drexler. Or it may fail. But Drexler's vision has caused many pioneers in the field of nanotechnology to stop and wonder what nanotechnology can be and what it can't be. By weeding through many diverse ideas, the good ideas stand out while the bad ideas fall by the wayside. This weeding process is one of the basic underpinnings of all science.

Whether or not the fantastic future envisioned by K. Eric Drexler becomes reality, one place where his vision of nanotechnology is flourishing is in science fiction. Since the early 1990s, science-fiction authors have picked up the idea of nanotechnology and used it as elements in their stories.

Neal Stephenson's novel *The Diamond Age* explores the idea of a society com-

Another vision of how the science of nanotechnology can help humankind. These nearly microscopic "robot worms" transform polluted soil into good, clean dirt ready for planting.

pletely dominated by Drexler's nanomachines. Every house is fitted with nanopipes that stream broken atoms into a device called a "compiler," which acts very much like Drexler's nanoassembler. The compiler makes any food you want. The air itself is filled with nanodevices. Some are almond-sized surveillance monitors, others are microscopic attack-and-defense nanorobots. One of the novel's main characters is extensively implanted with nanodevices that twitch his muscles constantly. This twitching, combined with a hormone pump, builds up his muscles without him actually having to do anything.

The Diamond Age and many other science-fiction stories have picked up the idea of nanotechnology and extended it into the far future. And yet, the science of nanotechnology has to date not produced a single useful device. Nanotechnology is still a science dominated by research papers and scientific meetings. What keeps researchers going is the dream and the promise of what the technology might become. And the stakes are high. If nanotechnology lives up to only half of its promise, there is hardly any aspect of society that will not be revolutionized by its innovations.

Smart Sensors

Science-fiction writers foresee a time when nanosized devices, inserted into the body, help enhance its performance. This picture shows an enhanced brain. Its central nanocomputer (in blue) stores vast amounts of data and releases it when necessary to surrounding brain cells.

One of the promises of nanotechnology is super-tiny computer chips that are inexpensive to make and require little power to operate. Although these nanoprocessors do not yet exist, scientists are already envisioning some of the ways nanoprocessors might be used. One of the chief applications would be as "smart sensors." Hospitals could be fitted with nanoprocessors to monitor important patient functions, such as heart rate and blood pressure. Aircraft could also benefit from these smart sensors in a similar fashion. Nearly all forms of manufacturing, from cars to books, would benefit from improved quality control. The nanoprocessors would not only be small, they would be powerful and smart as well, enabling them to make independent decisions and adjust things on their own.

G L O S S A R Y

antibodies Objects in the blood stream, such as white blood cells, that are used to fight disease and infection.

atomic force microscope (AFM) A device used to view and identify individual atoms by using a tiny probe in direct contact with the object.

bottom-up manufacturing An approach to manufacturing that begins with raw molecules and coaxes them to grow and self-assemble into useful components.

carbon nanotubes Collections of extremely tiny carbon molecules that shape themselves into pipelike structures.

classical physics Scientific theories that deal with matter and energy and govern the world we live in.

compiler A device to construct any object by selecting appropriate atoms and molecules and assembling them into the desired end product.

diodes Devices that channel the flow of electrons into one direction.

dip-pen lithography Nanofabrication technique using the atomic force microscope that works by dipping the tip of the AFM into a solution, bringing it close to a sheet of gold without actually touching the gold, and transferring molecules from the solution onto the gold through water that forms between the tip of the probe and the gold plate.

macroworld The world we see around us.

magnetotactic bacteria Tiny organisms that possess nanosized magnetic crystals that permit them to respond to Earth's magnetic field.

mesoscale A standard applied to the physics of the nanoworld.

microchip Integrated circuit that contains a complete central processing unit etched on a tiny piece of silicon.

microcircuits Very small patterns etched on a chip of silicon or other material.

micrometer Unit of measurement that equals one-millionth of a meter.

microorganism A living object of extremely small or microscopic size.

microprocessor chip The full name for a microchip.

millimeter Unit of measurement that equals one-thousandth of a meter.

Moore's Law Prediction made by Intel co-founder Gordon Moore that the number of transistors on integrated circuits would double every eighteen months.

motherboard Main circuit board of a computer.

nanoassembler A machine envisioned by scientist K. Eric Drexler for assembling atoms and molecules in new patterns.

nanocircuits The tiniest patterns of circuits etched on a microchip.

nanofabrication The production of ultra-small components, such as miniscule circuit boards.

nanomachines Working systems created on the nanoscale, usually measuring 100 nanometers or smaller.

nanometer Unit of measurement that equals one-billionth of a meter.

nanoprocessors Ultra-small devices to process data.

nanoshells Miniscule glass beads that are coated in a thin layer of gold and used to deliver molecules of medicine inside the body.

nanostructures Extremely tiny objects, built on the mesoscale.

nanosubmarines Minute devices envisioned as patrolling the human body and finding and eliminating abnormal conditions in the bloodstream and organs.

nanotechnology Deliberately designed structures that deal with minute objects in order to develop very tiny products.

nanotubes Extremely small pipelike structures made from carbon or other substances.

nanoworld The realm made up of the tiniest objects.

photolithography Technique to produce microchips that uses light to etch microprocessor patterns on a chip.

quantum physics Set of scientific theories that describes the behavior of individual atoms and the particles that make up atoms.

ribosomes Cellular structures where protein synthesis takes place.

scanning tunneling microscope (STM) A device that uses a small electric charge sent through an atomic object to provide an image of the object in a computer.

semiconducting transistor A switch that can be turned on or off between a conducting state and a nonconducting one.

smart sensors Devices that monitor functions and have the capability to make independent decisions and adjustments about those functions.

thiols Organic molecules that line up automatically in ordered patterns.

top-down manufacturing Technique that involves shrinking current manufacturing techniques to produce nanosized products.

tumor An abnormal growth or mass of tissue.

A

Air pollution, 42
Amplifiers, 38
Antibodies
  disease, identification of, 11
  function of, 8
  nanomagnets, use of, 10
  tumors, treatment of, 12
Artificial nanomagnet, 8
  virus detection, 10
Atomic force microscope
  (AFM), 30
  invention of, 33
Atoms and molecules
  atomic force microscope, use
    of, 32
  cell functions, 8
  creation of, 35
  macroworld, nanoworld com-
    pared, 19
  medicine, delivery of, 11
  nanoassembler handling, 41
  nanometer's size compared, 6
  silicon, replacement of, 34
  synthetic molecules, use of,
    15
  thiol solution, use of, 30, 32
  viewing devices for, 32, 33
  weighing of, 22

B

Bacteria
  disease, identification of, 10, 11
  magnetotactic, 8
  nanometer's size compared, 6
Blood test, 10
  genetic diseases, for, 15
Bottom-up approach, 35

C

Cancer treatment, 12, 13
  future of, 16
Carbon nanotubes, 35
  characteristics of, 38
  manufacturing of, 35
  problems with, 36
  thermal properties of, 40
Cell, 7
  bone damage, healing of, 18
  DNA, role of, 13
  energy sources for, 20
  inner structures of, 8
Columbia (space shuttle), 40
Compass (magnetic), 15
Compiler (science fiction), 44
Computers
  atomic force microscope,
    image from, 32, 33
  development of, 4
  memory, nanotechnology
    improving, 38
  microprocessors. See Micro-
    processors
  nanodevices compared, 23
  research, goals of, 34
  speed, need for, 24
  top-down technique, use of,
    38
  wiring, technology for, 39
Conductivity, 36
  manmade nanotubes, 38

D

Definitions
  diode, 35
  DNA, 13
  magnetotactic bacteria, 8

mask, 25
  nanostructure, 7
  quantum physics, 19
  transistor, 24
Diamond Age, The (Stephenson),
    44
Diode, 35, 38
Dip-pen lithography
  advantages of, 33
  described, 32
DNA, 7
  genetic disease, test for, 13
Drexler, K. Eric, 41
  criticism of, 42

E

Earth (planet), 8
Electricity, 33
  atomic force microscope, 32
  scanning tunneling micro-
    scope using, 33
  transistor, function of, 24
Electron, 34, 38
Electron beam lithography, 28
  atomic force microscope, use
    of, 32
Energy, 20
Environment, 15
Etching
  atomic force microscope, use
    of, 32
  electron beam, use of, 29
  light, use of, 25

G

Genetic code, 13
Germs. See Bacteria
Gold, 30

atomic force microscope
    using, 32
  nanoshells coating, 11, 13
Gravity, 8, 39

H

Hard drive (computer), 38
Human body
  abnormality, device detecting,
    16
  antibodies, role of, 8
  cancer treatment, efficacy of,
    12
  cell, structure of, 8
  DNA, role of, 13
  virus, detection of, 10

I

IBM (International Business
    Machines), 33

L

Laser, 25
Lenses, 25, 28
Library of Congress, 6
Lithography, 25
  X-rays, use of, 28

M

Macromachine, 42
Macroworld, 6, 19
Magnetic field, 8, 10
Magnetotactic bacteria, 8
Manufacturing
  bottom-up approach, 35
  microchips, of, 25
  quality control, improvement
    of, 45

robotic devices, 41
top-down methods, 28
Mask (computer). See Master
    pattern (computer)
Master pattern (computer), 25
dip-pen lithography, promise
    of, 32
electron beam, use of, 28
soft lithography, created with,
    30
X-rays, use of, 28
Matter (physics), 19
Medicine
bone damage, healing of, 15
cancer treatments, 12
genetic disease, test for, 13
identification of diseases, 11
magnetization of antibodies,
    8, 10
monitoring equipment, future
    of, 45
nanoshells, use of, 11
Mesoscale physics, 19
atom, weighing of, 22
power, sources of, 20, 23
Metal, 35, 38
circuit board, use in, 39
Metric system
DNA, measurements of, 7
mesoscale, 19
nanometric scale, 6
Microcircuits, 4, 24
Microprocessors
atomic force microscope, use
    of, 30
carbon nanotubes, use of, 35
electron beam, use of, 28

limitations of, 24
manufacturing process, 25
molecular transistors for, 34
Moore's Law, 27
sensors, future of, 45
size of, 25
soft lithography, use of, 30
transistor, function of, 24
X-rays, use of, 28
Molecules. See Atoms and mole-
    cules
Moore, Gordon, 27
Moore's Law, 27

N
Nanoshells, 11
cancer, treatment of, 13
Nanotubes. See Carbon nan-
    otubes
NASA, 40

O
Oxygen, 8
nanodevices, as energy for, 20

P
Photolithography, 24
described, 25
improvements in, 28
Photosynthesis, 20
Physics
mesoscale environment, 20,
    22, 23
quantum physics, 19
Pond, 8
Protein, 8

Q
Quantum physics, 19, 22

R
Radio tubes, 4
Random instability, 22
Reproduction (technology), 41
Ribosomes, 7
Robotic nanoassembler, 41
building of, 43
criticism with regard to, 42
Rubber stamp, 30

S
Scale (weighing), 22
Scanning tunneling microscope
    (STM), 33
Science fiction, 41, 43
Self-assembling product, 36, 37
experiments with, 38
robotic nanoassembler, 41
Sensor, 45
disease detection, 16
Silicon, 25, 30
carbon nanotubes as replace-
    ment, 35
dip-pen lithography, 33
molecular replacement for, 34
Soft lithography, 30
Solar energy, 20
Space exploration, 40
Stephenson, Neal, 44
Subatomic particles, 19

T
Technological advances
benefits of, 6

effects of, 4
nanoassemblers, 41
nanofabrication methods, 33
social issues, 44
top-down methods, benefits
    of, 28
Template (computer). See Mas-
    ter pattern (computer)
Thermal insulation, 40
Thiols, 30
atomic force microscope
    using, 32
Top-down technique (manufac-
    turing), 28, 38
Transistors, 24
carbon nanotubes, use of, 35,
    38
molecules, based on, 34
Moore's Law, 27

U
Ultraviolet beam
replacement of, 28
use of, 25

V
Virus, 8

W
Weighing devices, 22, 23

X
X-ray, 28

Z
Zyvex, 43